Blocks Of The Universe

AUDREY E. RANDLES

Copyright © 2012 Audrey Elizabeth Randles
All rights reserved.

Introduction

I would like to introduce the Theory of Matrix by rephrasing the words of Hermann Minkowski. The views of space and time, which I wish to introduce, have sprung from the soil of experimental psychology. 'Herein lies their strength. Their tendency is radical.'

The theory was born in the early 1990s. It was developed along with the Coresynthesis Psychological Model as a new theory of Self and Reality. The mathematical model of the Light Cone was an excellent starting point. Nevertheless, there was a space of ten years between the first explanations of the results and the association of these results with the characteristics mathematically introduced by Lorenz, Minkowski, and Einstein for the Light Cone.

We introduce a new understanding of the multimodal world and present the main principles associated with the multimodal Space-Time and energy structure of the objects and systems existing in the Universe.

Space is associated with the actuality of the period of time now occurring, and time is associated with the potentiality of the past, present, and future. The existing objects and systems and the Universe are the multimodal energy structures existing in space and time.

Background radiation, arranging the multimodal structures of the objects and systems, makes them the 'space-time-info-energy' building blocks of the Universe.

Cosmic background radiation, such as the cosmic microwave background radiation, and light carry the

'Genome' of the Universe and every existing object and system.

Stay well, and enjoy your reading.

Yours sincerely,
Audrey Elizabeth Randles

Image 1: 'NASA Telescope Spots Mystery in Fireworks Galaxy' Image credit: NASA/JPL-Caltech

Acknowledgement

We would like to express our gratitude to the National Aeronautics and Space Administration (NASA), NASA's Jet Propulsion Laboratory (JPL), and the California Institute of Technology (Caltech), for the impressive space images and exciting descriptions.

The views and opinions of the author, expressed in this book, do not necessarily state or reflect those of NASA's Jet Propulsion Laboratory, the California Institute of Technology or the National Aeronautics and Space Administration.

Contents

Introduction
Acknowledgement
Chapter 1 The Boundary
Chapter 2 The Universe Background
Chapter 3 Energy Draws Space and Time
Chapter 4 Space-Time Structures
Chapter 5 Blocks Of The Universe
Chapter 6 Dark Matter And Dark Energy
Chapter 7 The Centre Of The Universe
Chapter 8 Main Principles
Chapter 9 The Theory of Matrix Natural Laws
Chapter 10 Afterword
Content Use Policy

The Boundary

The human perception of time as a time-point, the current 'now', is the main boundary we operate in.

Consciously we operate 'now', at the '0' time-point. If you look inside your intelligent operating system, you notice that it is always 'now' for you. Our conscious 'I', the heart of our Individual Core, operates as a 'point', similar to a point in time - a moment.

Our momentary perception of time might be compared with the series of dots we perceive on the line made on one side of the 3-dimensional time cube of the 3-dimensional world. The human perception of time as a time-point is reflected in mathematics and other abstract sciences associated with numbers, quantities and space. It is influencing the abstract concepts, which are built on mathematics and applied to other disciplines, such as physics. Time characteristics in theoretical physics are mainly applicable in the settings of simultaneity and cause and effect replacing time dimensions. Some scientists offer the hypothesis of the time non-existence and replace time dimensions with 'point events'... reflecting time.

We build the cause and effects lines in our sciences and our lives, and we are still unsatisfied with predictions of future events. Without knowledge of time, we have chaos.

The above confirms the importance of time in our lives and our sciences, and our refusal to see the truth is the sublimation and repression of the time-related psychological conflict in men.

The theory of Matrix is a cosmological theory considering the new understanding of space and time, along with the Space-Time and energy structure of the existing objects and systems, including the Universe, visually 'empty' spaces, stars and galaxies, subatomic particles and holes, and other objects and systems existing in our dynamic world.

We analyse the main principles, associated with the multimodal structure of the objects and systems, and different modes of their functioning in different modalities of Space-Time.

Image 2: 'The Milky Way Centre Aglow with Dust' Image credit: NASA/JPL-Caltech

The Universe Background

There is no empty space in the Universe. The Universe is filled with energy. We perceive this energy in different forms, such as mass, kinetic energy, and other energy representations, acting in the volume of the Universe in the period of time now occurring. The latent, or potential, energy of the Universe and the existing objects and systems coded, coded and fixed by the latent, or potential, information, is associated with time. The latent, or potential, Info-Energy is the predominant form of energy in our Universe.

Time, space, mass, energy, and information do not exist independent of the objects and systems of the objects. They are the properties of the existing objects and systems, including planets, stars, systems holding Black Holes, subatomic particles and holes, visually 'empty' spaces, and other objects and systems existing in the Universe.

The Space-Time and Info-Energy (forms of energy) of the existing system with the qualities of the volume, mass, energy, and the time of existence are integrated into the system's multimodal structure, which we call the 'Matrix'.

Background radiation, including those currently known as cosmic background radiation, such as the cosmic microwave background radiation, infrared background, other forms of background radiation, and visible light arrange the Space-Time and energy structure of our dynamic world.

The 2-dimensional representations of background radiation arrange the latent, or potential, Info-Energy of the system into the 2-dimensional framework - the Info-Energy grid of the systems' multimodal Space-Time and Info-Energy structures. This 2-dimensional grid, arranging the Info-Energy of the system within the system's Space-Time, is the property of the system. It reflects the time, space, mass, energy, and information of the associated system.

The 2-dimensional representations of background radiation become dynamic in our dynamic world. These dynamic representations of background radiation build the internal framework and Space-Time and Info-Energy skeleton of every existing object and system in our dynamic world.

Any object, which is smaller than the wavelengths, supporting the Info-Energy grid of the object or the system, does not exist in Space-Time.

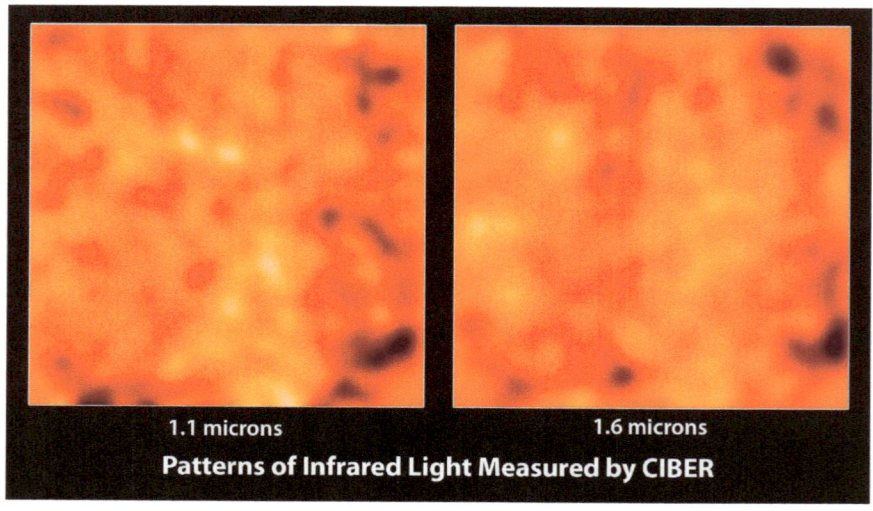

Image 3: 'Matching Patterns of Light' Image credit: NASA/JPL-Caltech

'These images from the Cosmic Infrared Background Experiment, or CIBER, show large patches of the sky at two different infrared wavelengths (1.1 microns and 1.6 microns)

after all known galaxies have been subtracted out and the images smoothed to enhance the large structures. CIBER sees similar patterns at different wavelengths, supporting the idea that the light patterns arise from the same source.' NASA

Background radiation, being the internal Space-Time and Info-Energy skeleton for the Universe and the existing objects and systems, arranges, supports, transports and transmits the Genome of the Universe in our dynamic world. Dynamic representations of background radiation carry the latent, or potential, Info-Energy structure of our dynamic world. It is the dynamic body of the multimodal Universe in our dynamic world.

'Observations fromNASA's Cosmic Infrared Background Experiment, or CIBER, have shown a surprising surplus of infrared light filling the spaces between galaxies.' NASA

According to the Theory of Matrix, CIBER is looking at the Genome of the Universe and every object and system, existing in the Universe. The genome of the Universe is the latent, or potential, Info-Energy of the Universe and objects and systems, existing in the Universe. The potential energy is coded and fixed by the potential information within the various representations of background radiation in the harmony of our dynamic world. The study of background radiation and its energy characteristics in the 2-dimensional Space-Time settings will allow to gain additional knowledge about the structure of the Universe and other existing objects and systems and provide a vast spectrum of opportunities for a broad range of the natural sciences.

Energy Draws Space and Time

Time is associated with latency and potentiality. Space is associated with the actuality of the current moment. Time and space do not exist independent of objects and systems. Time, space, mass, energy, and information as the properties of objects and systems are to be defined in relation to a frame of reference.

The system's time (or the object's time) is built by the latent, or potential, energy that is coded and fixed by the latent, or potential, information. We can compare this with a human genome that is the complete set of genetic information for a newborn child.

The volume of the system (or the object's volume) is filled with energy and associated information represented in mass, kinetic energy, and other energy representations acting in the volume of the system in the period of time now occurring.

Space influences the direction of the time 'flow'. The Space-Time and Info-Energy flow within the system is the system's response to the Space-Time and Info-Energy imbalance in the multimodal Space-Time and Info-Energy structure of the system. Gravity is a specific case of the system's Space-Time and Info-Energy imbalance. According to Albert Einstein,'the ponderable masses will be the determining factor in producing the field, or, according to the fundamental result of the special theory of relativity, the energy density...' [Albert Einstein, A Brief Outline of the Development of the Theory of Relativity (1921)].

According to the Theory of Matrix, the Matrix is not the field but the structure. The ponderable masses are the determining factor in producing the energy density, information density, and multimodal Space-Time and Info-Energy structures of the objects and systems.

The objects and systems of the objects of limited mass and energy are limited in space and time. Accordingly, the Matrixes for the objects and systems of limited mass and energy are limited in space and time.

The information does not exist independent of energy and matter. It depicts the forms of energy and matter. The information takes different forms and exists in the latent, or potential, and actual forms along with energy. Accordingly, the Matrixes for the objects and systems of limited mass and energy are limited in information. The density of information is proportional to the density of energy and matter, associated with the information.

The 2-dimensional representations of background radiation arrange the object or the system's 2-dimensional grid. The Matrix grid is a regular repeated, typically rectangular, grid-like 2-dimensional arrangement of the Info-Energy structure forming the object or the system's Matrix and keeping the object or the system's potential energy.

An example of the Matrix grid influence on the macro-scale is the subatomic and atomic processes associated with the regulation of heat in the body of the Universe, and the example of the grid-forming energy is the energy structure built by the cosmic microwave background radiation and infrared background (Image 1) carrying this heat and information associated with the regulation of heat in the Universe and objects and systems existing in the Universe.

Similarly, an example of the Matrix grid influence on the human scale is the chemical energy associated with the regulation of heat in the human body. An example of the grid-forming energy is the energy structure built by the energy of microwaves and infrared waves carrying this heat and information associated with the regulation of heat in the human body.

The existing objects and systems with qualities of mass, volume, energy, and time of existence have their multimodal Matrixes representing their total Info-Energy in space and time, including their past, present, and future. The Universe contains an infinity of the Matrixes, 'temporal fields', creating streams and rivers of space and time.

The object is a source of the Matrix, and, nevertheless, the Matrix is a self-organising system.

The Matrix grid is similar to a chromosome of living cells. A chromosome of living cells is carrying genetic information in the form of genes. The Matrix grid is a 2-dimensional Space-Time and Info-Energy framework carrying information in the form of Info-Energy blocks.

Matrixes are generated as an outcome of the binary operation of two Matrixes or by any of the following: reflection, self-duplication, binary fission, self-multiplication, and incorporation while maintaining anisotropy.

The description of the Matrix includes the typical characteristics introduced mathematically by Lorenz, Minkowski, and Einstein for the famous Light Cone, or the Matrix of Light, mainly associated with the Matrix symmetry, and some additional characteristics that are described in the Theory of Matrix.

The objects and systems, such as planets, stars, molecular clouds, galaxies and star clusters, visually 'empty' spaces, subatomic particles and holes, and other existing objects and systems, are enclosed at the centres of their Matrixes. They are tangled together by the Space-Time, Info-Energy, and Mass-Energy imbalance reflected in gravity, antigravity, electric charges, and other characteristics.

The Hawking radiation is a sign of the Matrix grid destruction at the Matrix centre or in the area of the Matrixes' connection if the '0' time-point centre was built, as a consequence of the tremendous excessive Info-Energy between connected Matrixes. Please see my book 'Black Holes and Supernovas' for details.

Background radiation provides a mechanism and conducts balancing Space-Time, Info-Energy, and Mass-Energy transformations, including those associated with gravity and antigravity.

Energies of background radiation and associated information, supporting the 2-dimensional grids of the multimodal Matrixes, make the multimodal objects and systems the STIE-blocks ('space-time-info-energy' building blocks) of the Universe.

Space-Time Structures

The object or the system's Matrix reflects the human perception of time as a point 'now', associated with the period of time now occurring, and current human understanding of two separate imperceptible periods of time - the past and the future.

The 2-dimensional grid, forming the Matrix, is predominantly the time-associated 2-dimensional Info-Energy structure. It forms the system's time and potential space underlying the system's volume. The system's volume is bounded by the 2-dimensional Info-Energy grid at the '0' time-point at the centre of its particular multimodal Matrix, representing the system's Space-Time, Info-Energy, and Mass-Energy.

The system's Space-Time

The system's Space-Time is represented in different modalities. Space-Time is represented in the Matrix as the Space of the Current Time associated with the period of time now occurring, Space of the Progressive Time associated with the future of the system, and Space of the Regressive Time associated with the system's past.

The Matrix Space of the Current Time is formed by the Matrix grid and connected with the Matrix Spaces of the Progressive and Regressive Time. The Space of the Current Time equals the volume of the system enclosed at the centre of the Matrix. The volume of the object is filled with the actual Info-Energy represented in mass, kinetic energy, and other energy representations, actually acting in the volume of the system in the period of time now occurring.

The Spaces of the Progressive and Regressive Time look like a hologram. They are immobile, unchanging, inaccessible, appeared empty and connected with the undifferentiated Continuum. The one-dimensional Space-Time Continuum is the one quality unified space-time-energy as the infinite duration. The Matrix Spaces of the Progressive and Regressive Time are symmetrical via the Matrix axis and 'o' time-point at the centre of the Matrix if the Matrix is balanced. Every Matrix demonstrates a tendency to obtain and retain a balance. There is no priority of the Progressive or Regressive Time in the Matrix. A mirror effect is possible.

The 'o' Space-Time point and the Arrows of Time

The 'o' time-point, reflecting the human perception of time as a point 'now', represents the time component of the 'o' Space-Time point. The system's volume is bounded by the Matrix grid at a 'o' time-point at the Matrix centre.

A 'o' Space-Time point, or 'a space-time-null-point', is a point 'here' and 'now'. The point 'here' and 'now' as a 'o' point was mentioned in Special Relativity in association with the mathematical model of the Light Cone. 'Let us call any world-point o as a space-time-null-point.' [Minkowski Hermann, 'Space and Time' (1920)]. 'A world-point is a 'here-now'. [Weyl Hermann, 'The Discussion concerning the Theory of relativity at the Meeting of Natural Scientists' (1920)]

The 'o' Space-Time point is the centre of the Matrix. It is a point of the Matrix symmetry. The Matrix Space-Time, Mass-Energy, and Info-Energy are symmetrical via the 'o' Space-Time point if the Matrix is balanced.

The 'o' Space-Time point is a point of Space-Time, Info-Energy, and Mass-Energy transformations, keeping space and time in a dynamic balance following the Space-Time, Mass-Energy, and Info-Energy Conservation Laws and Laws of Symmetry.

The 'o' Space-Time point of the Matrix may be represented as a coexistence of the 'o' time-point and 'o' space-point that are complementary, building the time system and the space system in the opposite direction.

The system's Space-Time Arrow is directed along the Space-Time axis.

According to the Theory of Special Relativity, the space axis of the Light Cone builds a perpendicular to the time axis. In compliance with our investigations of the Matrix and according to the Theory of Matrix, the Light Cone is an example of the Matrix for a flash of light. The Space-Time axis (other terms: Matrix axis, time axis, space axis) is the only axis of the Matrix.

The Space-Time Arrow, directed along the Space-Time axis, demonstrates a coexistence of four tendencies - the tendency of time and the contra-directed tendency of space, along with the tendency of the past and the contra-directed tendency of the future.

The Arrows of Time and contra-directed Arrows of Space reflect the time and space components of the Space-Time Arrow. They depict the Matrix tendency of time and the contra-directed tendency of space. Please see my book 'Space and Time' for details. The Arrows of Time are associated with the Spaces of the Progressive Time (the left cone associated with the future of the system on the Figures 1, 2) and Regressive Time (the right cone associated with the past of the system on the Figures 1, 2). They are directed along the Matrix axis. The time component of the Space-Time Arrow is represented as the Arrow of the Progressive Time and Arrow of the Regressive Time. The Arrows of the Progressive and Regressive Time indicate the direction of the Space-Time and energy 'flow' within the multimodal structure of an object or a system. The direction of the Arrows of Time provides us with an opportunity to detect the possible changes of the Space-Time direction and the reverse of the Space-Time and energy flow within the objects and systems, for example, systems holding Black Holes, the Sun, our planet, and the Universe.

TRM and SRM

Two types of the multimodal structures have been identified - the Time-Rising Matrix, or TRM (Figure 1), and Space-Rising Matrix, or SRM (Figure 2).

The Time-Rising Matrix, or the TRM, is the multimodal structure of a radiating object or a system, such as high energy massive systems, stars and planets, radiating galaxies and star clusters tangled by gravity.

The Space-Rising Matrix, or the SRM, is the multimodal structure of the large spacious and Space-Rising objects and systems, including our Universe and systems creating Black Holes, developing surface antigravity.

The direction of the Arrows of Time in the TRMs (Figure 1) is different from the direction of the Arrows of Time in the SRMs (Figure 2).

The Arrows of Time are drawn as the yellow arrows on the reproduction of the Matrixes, while the direction of the Space of the Current Time is represented by the blue arrows, situated following the human perception of the systems as exhibiting three space dimensions (x, y, and z, or a combination of three directions, which can be chosen from the terms: length, width, height, depth, and breadth).

The TRMs of the radiating objects and systems are formed by the 2-dimensional grid. They have a form of Riemannian Manifold with the negative curvature (Figure 1).

In the TRM, the Space of the Progressive Time is a deviation from the Progressive Time with the characteristics of the time 'flow', and the Arrow of the Progressive Time is directed from the '0' time-point at the centre of the TRM to the future of the associated system.

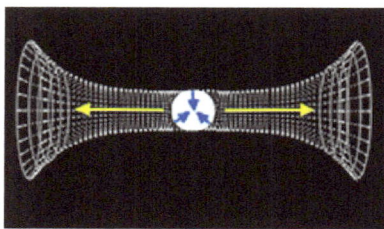

Figure 1: The Time Rising Matrix

The Space of the Regressive Time is a deviation from the Regressive Time with the characteristics of the time 'contra-flow', and the Arrow of the Regressive Time is directed from the '0' time-point to the past of the associated system.

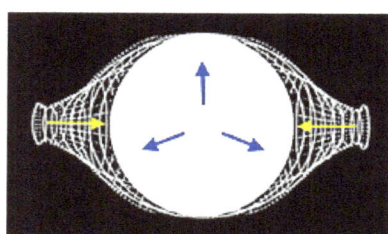

Figure 2: Space-Rising Matrix

The SRMs of the Space-Rising objects and systems, including our Universe and systems creating Black Holes, have a form of Riemannian Manifold with the positive curvature (Figure 2).

In the SRM, the Space of the Progressive Time is a deviation from the Progressive Time with the characteristics of the time 'contra-flow', and the Arrow of the Progressive Time is directed from the future of the system to the '0' time-point at the Matrix centre.

The Space of the Regressive Time is a deviation from the Regressive Time with the characteristics of the time 'flow', and the Arrow of the Regressive Time is directed from the past of the system to the '0' time-point at the Matrix centre.

The Arrows of Time are symmetrical if the Matrix is balanced in Space-Time, Mass-Energy, and Info-Energy.

Strictly speaking, the reproduction of the Matrix grid on the paper is not exact though the form of the Matrix is correct. It is related to the difficulties to reproduce the 2-dimensional structure on the paper. I want to emphasise that the Matrix grid, forming the Matrix Spaces of Time, is a rectangular, regular repeated, with the equal distances, 2-dimensional infinitely thin filament containing the system's latent, or potential, energy and related information depicting the form of energy.

Blocks Of The Universe

According to the theory of General Relativity, energy curves Space-Time [Albert Einstein, 'Relativity: The Special and General Theory' (1916)]. The Theory of Matrix supports this concept.

The Matrix has a form of the Riemannian Manifold with the positive or negative curvature. A curvature of the Matrix depends on the Space-Time, Mass-Energy, and Info-Energy properties of the associated object/system.

The distances along paths on the surface of the Matrix and angles are to be measured as the characteristics of the latent, or potential, Info-Energy of the system associated with the Matrix 2-dimensional grid. The notion of the curvature is to be defined in a way that is intrinsic to the manifold.

The curvature of the Matrix grid depends on the Space-Time, Info-Energy, and Mass-Energy properties of the system. The curvature of the Matrix does not depend on how the surface might be enclosed in 3-dimensional or higher-dimensional spaces. The curvature of the Matrix does not depend on how the 3-dimensional space is enclosed at the centre of the Matrix.

Black pixels, reading zero (Figures 1 and 2), corresponding to the potential information, are fixed by the potential energy of the 2-dimensional grid. They are associated with the timing mechanisms and sweep rates, and the address of a pixel corresponds to its Space-Time coordinates.

We suppose that the blocks of information, which are involved in the actuality or located relatively close to the point 'now' in time, are situated closer to the centre of the Matrix, then other latent information. Hypothetically, if the blocks of information are associated with the Matrix Space of the Current Time at the centre of the Matrix, then they might have a 3-dimensional structure.

Two opposite vector-forces, the Matrix grid vector-force of pressure and the object/system's vector-force of resistance, act in a dynamic balance in the Space of the Current Time of the Matrix centre. They influence Space-Time, Mass-Energy, and Info-Energy transformations providing an unbalanced object/system with an opportunity to establish a balance.

The potential and actual Info-Energy structures counteract and keep a balance in the Matrix, retaining total Space-Time, Mass-Energy, and Info-Energy of the Matrix unchanged.

The high energy massive radiating objects and systems and Space-Rising non-radiating objects and systems exhibit the Space-Time imbalance and associated Mass-Energy and Info-Energy imbalance and, accordingly, the balancing Space-Time, Mass-Energy, and Info-Energy transformations. Space-Time transformations are the function of energy. Transformations of the actual information into the potential form proceed along with energy transformations from the actual form into the potential form. This process may be reversed.

The frequency of the electromagnetic waves and their wavelength influence the Matrix grid, changing Space-Time properties via the energy transfer and retaining the Space-Time, Info-Energy and Mass-Energy Conservation Laws and

Matrix Laws of Symmetry. Background radiation provides a mechanism for the Space-Time, Info-Energy, and Mass-Energy transformations and transmissions in our dynamic world.

The Space-Time imbalance and associated influence of the Matrix forces are reflected in the qualities of gravity, antigravity, and specific rotation of the volume of these multimodal objects and systems about their Space-Time axis.

Please consider the qualities of gravity, antigravity. Gravity, along with gravitational acceleration and space and mass formation, is prompted by the peripheral areas of the unbalanced high energy massive radiating objects and systems, such as our planet, and central areas of the unbalanced Space-Rising non-radiating objects and systems, such as the systems holding Black Holes. Antigravity, along with anti-gravitational deceleration and space and mass degeneration, is prompted by the central areas of the unbalanced high energy massive radiating objects and systems and peripheral areas of the unbalanced Space-Rising non-radiating objects and systems. The balancing transformations are reflected in the gravity and antigravity propagation, and space and mass formation and degeneration within our dynamic world. Dynamics preserve the Laws of Space-Time, Mass-Energy and Info-Energy Reversibility, Conservation, Transformation, Limitation, and Symmetry. These Natural Laws are responsible for the harmony of the existing Universe.

The rotation of the multimodal systems, including the unbalanced high energy massive radiating objects and systems and Space-Rising non-radiating objects and systems, is associated with their specific geometry and the influence of the Matrix forces. The associated rotation of the 2-dimensional grid is represented in the rotation of background radiation.

Anisotropies, or irregularities, of background radiation, such as the irregularities of the cosmic microwave background radiation, are associated with the two opposite forces, acting at the centre of the Matrix of the Universe and Matrixes of the existing objects and systems. Irregularities of background radiation are related to the process of the Space-Time, Info-Energy, and Mass-Energy transformations, generation of new objects and systems and their multimodal structures, incorporation and absorption of the Matrixes and associated objects, systems, and visually empty spaces existing in our dynamic world.

Dark Matter and Dark Energy

The Space-Time imbalance, such as the excessive time and space deficit, and associated Info-Energy and Mass-Energy imbalance reflect the qualities of gravity prompted by the central areas of the large spacious non-radiating objects and systems and the peripheral areas of the high energy massive radiating objects and systems. The Space-Time, Info-Energy, and Mass-Energy transformations are reflected in gravitational acceleration, space, energy, and matter generation within the system.

The Space-Time imbalance, such as the excessive space and time deficit, and associated Info-Energy and Mass-Energy imbalance reflect the qualities of antigravity prompted by the central areas of the high energy massive radiating objects and systems and the peripheral areas of the large spacious non-radiating objects and systems. The Space-Time, Info-Energy, and Mass-Energy transformation are reflected in anti-gravitational deceleration, space, energy, and matter degeneration and reduction within the system.

Space and mass degeneration and reduction within the Universe are balanced with space and mass generation prompted by the central areas of the Universe. Space and mass formation, prompted by the central areas of the Universe, coexists in a dynamic balance with space and mass degeneration prompted by its periphery.

Antigravity is most intense in the peripheral areas of the Universe. Anti-gravitational processes, prompted by the periphery of the Universe, are related to space and mass degeneration and reduction within some regions of the Universe.

Space, energy, and matter degenerate throughout the system into different forms of non-radiating matter and energy, mainly associated with anti-gravitational processes in the area and reflected in the existing 'Dark matter' and 'Dark energy', and the regions of inflation within the Universe.

'Dark matter is non-luminous matter which is postulated to exist in space and which could take either of two forms: weakly interacting particles (cold dark matter) or high-energy randomly moving particles created soon after the Big Bang (hot dark matter).' (Oxford dictionary)

Image 4: 'The Clumping Behavior of Galaxies' Image credit: NASA/JPL-Caltech

'Active, supermassive black holes at the hearts of galaxies tend to fall into two categories: those that are hidden by dust, and those that are exposed. Data from NASA's Wide-field Infrared Survey Explorer, or WISE, have shown that galaxies with hidden supermassive black holes tend to *clump together in space more than the galaxies with exposed, or unobscured, black holes.*

'This enhanced image shows galaxies clumped together in the Fornax cluster, located 60 million light-years from Earth. The picture was taken by WISE, but has been artistically enhanced to illustrate the idea that clumped galaxies will, on average, be surrounded by larger halos of dark matter (represented in purple). Because dark matter, like normal matter, has gravity, it will pull galaxies toward it, causing them to clump.

'Astronomers don't know why the hidden black holes would have larger halos of dark matter, but are intrigued by the surprising finding and are investigating further.'
NASA

Dark energy is a theoretical form of energy postulated to act in opposition to gravity and to occupy the entire Universe, accounting for most of the energy in it and causing its expansion to accelerate.

Under the Theory of Matrix, the presence of the dark matter, dark energy, and other forms of non-radiating matter indicate the location of the space, energy and mass degeneration and reduction, along with the anti-gravitational processes in the area. We suppose we could meet the specific forms of the degenerated matter and energy associated with anti-gravitational processes in the central regions of our planet, our Sun, and other massive radiating objects and systems, such as planets, stars, and radiating galaxies tangled by gravity.

The projection of the different representations of the 'o' time-point, existing in the TRMs' 2-dimensional time settings, into the volume of the high energy massive radiating objects and systems, such as our planet, might give us an exciting Space-Time region located within the projection in our dynamic world. Please see my book 'Space and Time' for more details of the Space-Time structure of the existing objects and systems.

We suppose these regions develop antigravity along with space and mass degeneration. Perhaps the elements of the degenerated matter and energy are not so far from us as we think.

We call the degenerated matter and energy, associated with anti-gravitational processes within the central areas of the high energy massive radiating objects and systems, 'black matter' and 'black energy' to express the difference between the degenerated matter and energy associated with anti-gravitational processes within the radiating objects and systems and anti-gravitational processes within the non-radiating objects and systems.

The influence of the Matrix forces causes the balancing Space-Time, Info-Energy, and Mass-Energy transformations.

The gravitational and anti-gravitational horizon is the object/system's 2-dimensional Info-Energy grid, processing Space-Time, Info-Energy, and Mass-Energy transformations.

Dynamic representations of background radiation arrange, support, transport, and transmit the Space-Time, Info-Energy and Mass-Energy transformations, including those reflected in gravitational acceleration and anti-gravitational deceleration, throughout our dynamic world.

Gravitational processes in the central areas of the Universe and anti-gravitational processes in its peripheral areas do not contradict but prompt the local processes associated with the formation and degeneration of spaces and masses throughout the Universe.

The Centre Of The Universe

Gravity is most intense at the centre of the Universe. Gravitational processes, prompted by the centre of the Universe, are related to the generation of spaces and masses within some regions of the Universe.

Considering a direction to the centre of the Universe, we can point on the temperature fluctuations of the ancient light.

According to the Theory of Matrix, the asymmetry in the temperature fluctuations of the cosmic microwave background across two halves of our sky (with the extent of these variations more significant on the hemisphere shown at right than the one at left) indicates the central areas of the Space-Rising non-radiating system. It possibly indicates the centre of the Universe with the intense formation of the masses and kinetic energy and, accordingly, the greater extent of the temperature variations on the hemisphere shown at the right of the Planck mission image (not in this book).

The greater extent of the temperature variations is reflected in fluctuations of the cosmic microwave background radiation.

The central areas of the Universe bring about Space-Time transformations and associated Info-Energy and Mass-Energy transformations reflected in the formation of spaces, masses, and kinetic energy, filling up and acting in the volume of the Universe in the period of time now occurring.

The associated Space-Time, Info-Energy, and Mass-Energy transformations are processed and transmitted by different forms of background radiation.

Humans experience qualities associated with the intermediate region of the Universe and local gravitational processes are perceived, for example, as gravitational acceleration on the surface of our planet.

Please see more details about the structure of the Universe in my book 'Matrix of the Universe.

Image 5: 'Scene of Multiple Explosions' Image credit: NASA/JPL-Caltech

'This composite image shows Z Camelopardalis, or Z Cam, a double-star system featuring a collapsed, dead star, called a white dwarf, and a companion star, as well as a ghostly shell around the system. The massive shell provides evidence of lingering material ejected during and swept up by a powerful classical nova explosion that occurred probably a few thousand years ago.

'Z Cam was one of the first known recurrent dwarf nova, meaning it erupts in a series of small, "hiccup-like" blasts, unlike classical novae, which undergo a massive explosion. That's why the huge shell around Z Cam caught the eye of astronomer Dr. Mark Seibert of Carnegie Institution of Washington in Pasadena, Calif. - it could only be explained as the remnant of a full-blown classical nova explosion. This finding provides the first evidence that some binary systems undergo both types of explosions. Previously, a link between the two types of novae had been predicted, but there was no evidence to support the theory.' NASA

Main Principles

The main principles and natural laws of the Theory of Matrix are related to natural phenomena and applicable to the Grand Universe, Multiverse, our Universe, and every existing object and system of the objects.

The main principles of the Theory of Matrix:

1. The concept of relativity and Einstein's Special Theory of Relativity state that all motion is relative and that the velocity of light in a vacuum has a constant value which nothing can exceed.

The Theory of Matrix adds that all dynamics in our dynamic world is associated with the difference in our perception of space and time. The relative difference in our perception of space and time makes this world dynamic for us - the relation between 1-dimensional space and 1-dimensional time we perceive as speed.

The numerical value of the velocity of light in a vacuum is the Space-Time Coefficient, which defines the relation between 1-dimensional space and 1-dimensional time. The Space-Time Coefficient [c] equals the numerical value of the speed of electromagnetic waves propagation in a vacuum.

The value of the speed of light in a vacuum (c) is 299 792 458 m s-1 (meter per second). It is a CODATA Fundamental Physical Constant.

2. The Coefficient of Transformation $[c]^2$ applies to the decisions associated with the Space-Time, Mass-Energy, and Info-Energy Transformations, including those reflected in gravity and antigravity propagation. The Coefficient of

Transformation [c]² equals the squared numerical value of the speed of electromagnetic waves propagation in a vacuum.

3. Time, space, mass, energy, and information do not exist independent of the objects and systems. They are the properties of the objects and systems. Time, space, mass, energy, and related information are to be defined in relation to a frame of reference.

4. Every existing object/system with qualities of mass, volume, energy, and time of existence has its multimodal Matrix, representing the object/system's Space-Time, Info-Energy, and Mass-Energy characteristics.

The multimodal objects and systems carry the properties of different Space-Time modalities. Every existing single object and every system of the objects, including visually 'empty' spaces, subatomic particles and holes, quanta, humans, systems holding Black Holes, other objects and systems, and our Universe, are the multimodal objects and systems. The characteristics of an object or a system are different in different modalities of Space-Time. The system's multimodal Matrix represents the system's multimodal Space-Time and Info-Energy structure.

The 1-dimensional Space-Time is the one quality unified space-time-energy as the infinite duration. It is represented in the 1-dimensional Space-Time Continuum and associated with the system's past and future.

The fundamental 2-dimensional Space-Time is built by the latent, or potential, Info-Energy of the system. It underlies the volume of the system filled with the actual Info-Energy in the period of time now occurring.

The speed of light and the Planck units, such as Planck length, Planck time, and Planck energy, represent the limits of Space-Time and Info-Energy imbalance and, accordingly, the limits of the object or the system's existence. They initiate the reverse of the Space-Time and Info-Energy flow in the multimodal structure of the object or the system.

5. Space-Time Equivalence

Albert Einstein has specified the relation between mass and energy of the object. 'If an amount of energy E be given to a body, the inertial mass of the body increases by an amount E/c^2, where c is the velocity of light in vacuo. On the other hand, a body of mass m is to be regarded as a store of energy of magnitude mc^2.' [Einstein Albert, A Brief Outline of the Development of the Theory of Relativity (1921)].

Please consider the equation for Mass-Energy equivalence $E = mc^2$ with respect to the 2-dimensional Space-Time understanding.

In the Theory of Matrix, we develop the Theory of Light, uncover unusual properties of light, specify the relation between space and time of the object/system, and introduce Space-Time equivalence.

$$E \div [c]^2 m = l^2 \div t^2 \,,$$

where E is energy, m is mass, l^2 is the 2-dimensional space, t^2 is the 2-dimensional time, and $[c]^2$ is the Coefficient of Transformation.

The Coefficient of Transformation $[c]^2$ equals the squared numerical value of the speed of electromagnetic waves propagation in a vacuum, E is energy, m is mass.

'Relativity theory ... shares with the corpuscular theory of light the unusual property that light carries inertial mass from the emitting to the absorbing object.' [Einstein Albert, The Development of Our Views on the Composition and Essence of Radiation (1909)].

We specify the relation between space and time.

The universal proportionality exists between equivalent amounts of the fundamental 2-dimensional time and the fundamental 2-dimensional space, corrected with the squared numerical value of the speed of electromagnetic waves propagation in a vacuum.

Energy is associated with 2-dimensional time the same way as mass, corrected with the squared numerical value of the speed of electromagnetic waves propagation in a vacuum, is associated with 2-dimensional space.

$Et^2 = [c]^2 m l^2$,

where E is energy, m is mass, l^2 is the 2-dimensional space, t^2 is the 2-dimensional time, and $[c]^2$ is the Coefficient of Transformation.

Rest mass and rest energy remain proportional to one another the same way as the 2-dimensional space and the 2-dimensional time remain proportional to one another.

The universal proportionality factor, taken as the squared numerical value of the speed of electromagnetic waves propagation in a vacuum, is the Coefficient for Space-Time, Mass-Energy, and Info-Energy Transformations.

6. We state the principle of Space-Time conservation: the volume of a system may change, although the total Space-Time of the system remains constant.

We reaffirm the principle of conservation of energy: the total mass of a system may change, although the total energy of the system remains constant.

Space-Time, Mass-Energy, and Info-Energy of the system are conserved in the system's Matrix.

Some amount of time might be transformed into the proportional amount of space, and some amount of space may be transformed into the proportional amount of time. Space-Time transformations are the function of energy.

Some amount of mass might be transformed into the proportional amount of energy, and some amount of energy may be transformed into the proportional amount of mass.

7. The 2-dimensional representations of background radiation, arranging the Matrix external framework of the Universe and objects and systems existing in the Universe, build simultaneously the internal framework and Space-Time and Info-Energy skeleton of every object and system existing in our dynamic world. Energies of background radiation and associated information make multimodal Matrixes the STIE-blocks ('space-time-info-energy' building blocks) of the Universe.

8. We support and develop the Equivalence principle of General Relativity. Equivalence principle of General Relativity states that at any point of Space-Time the effects of a gravitational field cannot be experimentally distinguished from those due to an accelerated frame of reference.

The Theory of Matrix presents the mentioned gravitational field as the specific Space-Time imbalance, such as the excessive time and deficit of space, supported by the associated Info-Energy and Mass-Energy imbalance. The effects of this specific Space-Time imbalance in an accelerated frame of reference cannot be experimentally distinguished from those due to gravity.

The specific Space-Time imbalance, such as the excessive space and deficit of time, is reflected in antigravity. The effects of this specific Space-Time imbalance in a decelerated frame of reference cannot be experimentally distinguished from those due to antigravity.

9. Equilibrium of the Matrix is reached if the Matrix and associated object/system are balanced in energy, associated information, and Space-Time. Internal equilibrium of the

Matrix occurs when the Matrix resultant force is zero. The Matrix resultant force equals zero if the Matrix is balanced and symmetrical via the '0' Space-Time point and via the Matrix axis.

In case of the contact equilibrium between Matrixes, Space-Time, Info-Energy, and Mass-Energy are transferred through the contact paths, and the relation of the equilibrium is transitive, reflexive, and symmetrical. The contact equilibrium is reached if the system is balanced.

10. Internal thermal equilibrium of a Matrix is reached if the Matrix is balanced that means that no heat enters or leaves it, it is balanced under its intrinsic characteristics, symmetrical, and its temperature is spatially and temporally uniform.

In case of the contact thermal equilibrium between Matrixes, heat is transferred through the contact paths, and the relation of thermal equilibrium is transitive, reflexive, and symmetrical. The contact equilibrium is reached if the system is balanced.

The principles of the Theory of Matrix will be reappraised in the following books.

The Theory of Matrix Natural Laws

The Theory of Matrix declares the Laws of Space-Time, Info-Energy, and Mass-Energy Reversibility, Transformation, Conservation, Limitation, and Symmetry, which apply to the Universe and the existing objects and systems.

1. The Laws of Symmetry is based in the Newton's First Law (the Law of Inertia).

Every multimodal object/system tends to obtain and retain its balance and symmetry via the '0' Space-Time point and via the Space-Time axis. A mirror effect is possible. The Matrix Space-Time, Info-Energy, and Mass-Energy are symmetrical via the '0' Space-Time point and via the Space-Time axis if the Matrix is balanced.

2. The Limitation Laws

The objects and systems of the objects, limited in space, are limited in time. The objects and systems of the objects, limited in time, are limited in space. The Matrixes for these objects and systems are limited in Space-Time. The objects and systems of the objects of limited mass, energy, and associated information are limited in space and time. The objects and systems, limited in space and time, are limited in mass, energy and associated information. The Matrixes for the objects and systems of limited mass, energy, and associated information are limited in Space-Time. The Limitation Laws apply to every existing object/system and every Matrix.

3. The Laws of Space-Time, Info-Energy, Mass-Energy Conservation

The total Space-Time of every existing object/system equals the total Space-Time that this object/system possesses for the total period of its existence, including its past, present, and future. The total Space-Time of every existing object/system is conserved in the object/system's Matrix. The volume of the object/system may change, although the total Space-Time of the object/system remains constant. The time of the object/system may change, although the total Space-Time of the object/system remains constant.

The total energy and associated information of every existing object/system equal the total Info-Energy that this object/system possesses for the total period of its existence, including its past, present, and future. The total Info-Energy of every existing object/system is conserved in the object/system's Matrix. The actual Info-Energy of an object/system may change, although the total Info-Energy of the object/system remains constant. The latent Info-Energy of an object/system may change, although the total Info-Energy of the object/system remains constant. The Law of Info-Energy Conservation is a consequence of the Law of Mass-Energy Conservation.

The total Mass-Energy of every existing object/system equals the total Mass-Energy that this object/system possesses for the total period of its existence, including its past, present, and future. The total Mass-Energy of every existing object/system is conserved in the object/system's Matrix. The total mass of an object/system may change, although the total Mass-Energy of the object/system remains constant.

4. The Laws of Space-Time, Info-Energy, Mass-Energy Reversibility

The Space-Time of the object/system may be reversed. The Arrows of Time of the object/system may be reversed. The Space-Time of the object/system's Matrix may be reversed. The Space-Time Reversibility is determined by the Space-Time, Info-Energy, and Mass-Energy resources of the object/system.

The Info-Energy of the object/system may be reversed. The Info-Energy of the object/system's Matrix may be reversed. The Info-Energy Reversibility is determined by the Space-Time, Info-Energy, and Mass-Energy resources of the object/system.

The Mass-Energy of the object/system may be reversed. The Mass-Energy of the object/system's Matrix may be reversed. The Mass-Energy Reversibility is determined by the Space-Time, Info-Energy, and Mass-Energy resources of the object/system.

The Space-Time, Info-Energy, and Mass-Energy of the object/system's Matrix may be reversed and, accordingly, the Matrix may be reversed. The Matrix Reversibility is determined by the Space-Time, Info-Energy, and Mass-Energy resources of the object/system.

The Laws of Space-Time, Info-Energy, Mass-Energy Reversibility apply to every existing object/system and every Matrix.

The Laws of Space-Time, Info-Energy, Mass-Energy Reversibility are ruled by the Laws of Space-Time, Info-Energy, and Mass-Energy Conservation, Transformation, Limitation, and Symmetry.

5. The Laws of Space-Time, Info-Energy, Mass-Energy Transformations

Some amount of time might be transformed into the proportional amount of space, and some amount of space may be transformed into the proportional amount of time. The Space-Time transformation is a function of energy.

Some amount of latent Info-Energy might be transformed into the proportional amount of actual Info-Energy, and some amount of actual Info-Energy may be transformed into the proportional amount of latent Info-Energy.

Some amount of energy might be transformed into the proportional amount of mass, and some amount of mass may be transformed into the proportional amount of energy.

The Laws of Space-Time, Info-Energy, Mass-Energy Transformations apply to every existing object/system and every Matrix. The Laws of Space-Time, Info-Energy, Mass-Energy Transformations are ruled by the Laws of Space-Time, Info-Energy, and Mass-Energy Conservation, Reversibility, Limitation, and Symmetry.

The Coefficient of Transformation $[c]^2$ applies to the decisions associated with Space-Time, Info-Energy, Mass-Energy Transformations, including those associated with gravity and antigravity propagation. The Coefficient of Transformation $[c]^2$ equals the squared numerical value of the speed of electromagnetic waves propagation in a vacuum. The Space-Time Coefficient $[c]$ applies to the transmission of the Space-Time, Info-Energy, Mass-Energy Transformations in our dynamic world. The Space-Time Coefficient $[c]$ equals the numerical value of the speed of electromagnetic waves propagation in a vacuum.

The main principles and natural laws of the Theory of Matrix are related to natural phenomena. They apply to the Grand Universe, Multiverse, our Universe, and every existing object and system of the objects.

Afterword

The Theory of Matrix is a new theory combining the elements of psychology, cosmology, and astrophysics.

Dr Randles developed the program for psychological investigations of the Universal Matrix along with the development of the Coresynthesis Psychological Model in early 1990s.

The Theory of Matrix was introduced by Dr Audrey E. Randles in her work 'The Theory of Matrix' in 2012. When Dr Randles published her work, the Light Cone was still considered a specific case applicable to the flash of light. She brought forward the idea of the universality of the Space-Time structure of the objects and a new vision on Space-Time physics.

Following the analysis of the parallels and variations between the Universal Matrix and Light Cone Dr Randles calls the Light Cone 'the Matrix of Light' and presents the Theory of Matrix as the relative importance for the true understanding of the multimodal world.

The Theory of Matrix introduces the new vision of the multidimensional world with respect to multidimensional time. Time is no longer seen as another dimension of space, nor as a momentary feature of an event but as a multidimensional element in its own right.

Dr Randles associated space with the Actuality and time with the Potentiality or Latency. Therefore, the objects are viewed as the multimodal, multidimensional objects in Space-Time.

Books on Cosmology by Audrey E. Randles:
'Systems Theory in Cosmology' (2020)
'The Multimodal World' (2020)
'Black Holes and Supernovas' (2016)
'Grand Universe' (2016)
'Antigravity' (2015)
'Supernovas' (2015)
'The Primary Black Hole of the Universe' (2015)
'Energy in Cosmology' (2014)
'Gravity and Antigravity to the Point' (2014)
The Theory of Matrix series of books (2012 - 2013):
'Blocks of the Universe'
'Space and Time'
'Energy of Existence'
'Gravity and Rotation'
'Black Holes'
'Matrix of the Universe'
 The new 2020 books on Cosmology include Kindle ebook and a paperback of the same title at Amazon's Book Store.

Content Use Policy

Content may be used for any purpose without prior permission, subject to the special cases noted below.

By downloading the material, the user agrees:

1. to use a credit line in connection with the content. Unless otherwise noted in the caption information for any content and images the credit line should be

"Audrey E. Randles, 'The Theory of Matrix. Blocks of the Universe' (2013) red 2020".

2. that we do not represent others who may claim to be authors or owners of copyright of any of the content, and make no warranties as to the quality of the content;

3. that we shall not be responsible for any loss or expenses resulting from the use of the content, and you release and hold us harmless from all liability arising from such use.

Special Cases:

This content is available for educational, journalistic, personal uses and scientific research following a scientific code of ethics.

Restrictions are placed on commercial uses. To obtain permission for commercial use, contact the copyright owner Dr Audrey E. Randles.

www.ingramcontent.com/pod-product-compliance
Lightning Source LLC
Chambersburg PA
CBHW040249220526
45473CB00001B/419